MEASURE IT!

RACHEL FIRST

Consulting Editor, Diane Craig, M.A./Reading Specialist

Sandcastle

An Imprint of Abdo Publishing
abdopublishing.com

abdopublishing.com

Published by Abdo Publishing, a division of ABDO, PO Box 398166, Minneapolis, Minnesota 55439.

Printed in the United States of America, North Mankato, Minnesota

102015
012016

THIS BOOK CONTAINS RECYCLED MATERIALS

Editor: Liz Salzmann
Content Developer: Nancy Tuminelly
Cover and Interior Design and Production: Mighty Media, Inc.
Photo Credits: Shutterstock, Wikimedia Commons

Library of Congress Cataloging-in-Publication Data

First, Rachel, author.
Measure it! : fun with length & distance / Rachel First ; consulting editor, Diane Craig, M.A./ reading specialist.
pages cm -- (Math beginnings)
ISBN 978-1-62403-934-8
1. Measurement--Juvenile literature. 2. Length measurement--Juvenile literature. I. Title.
QA465.F47 2016
530.8--dc23
2015020585

SandCastle™ Level: Transitional

SandCastle™ books are created by a team of professional educators, reading specialists, and content developers around five essential components—phonemic awareness, phonics, vocabulary, text comprehension, and fluency—to assist young readers as they develop reading skills and strategies and increase their general knowledge. All books are written, reviewed, and leveled for guided reading, early reading intervention, and Accelerated Reader™ programs for use in shared, guided, and independent reading and writing activities to support a balanced approach to literacy instruction. The SandCastle™ series has four levels that correspond to early literacy development. The levels are provided to help teachers and parents select appropriate books for young readers.

EMERGING • BEGINNING • **TRANSITIONAL** • FLUENT

[redacted]

Contents

LONG and FAR

It's fun to measure things!

Measuring tells us length.

Length is how long something is.

Jeff measures his foot.

Measuring also tells us distance. Distance is how far **apart** things are.

Eli loves the ocean.

He wonders how far away it is.

Which SYSTEM?

Two systems measure length.

METRIC SYSTEM	STANDARD SYSTEM
millimeter	inch
centimeter	foot
meter	yard
kilometer	mile

Most countries use the metric system. The United States uses the standard system.

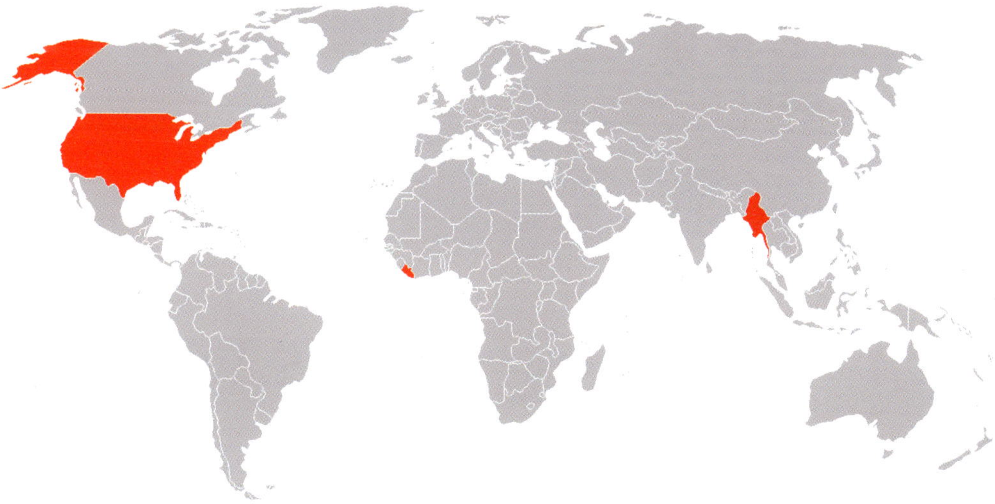

— Countries that use the metric system

— Countries that use the standard system

Little LENGTHS

Centimeters are small. This line is 1 centimeter.

CENTIMETER

Inches are longer. But they are still small. This line is 1 inch.

INCH

Try It!

Find some small things. Try rocks, nuts, or anything! Measure them! Use the lines on the opposite page. Which are about 1 centimeter? Which are about 1 inch?

RULERS RULE!

Rulers measure length. Rulers are 12 inches long. This is called 1 foot.

Dean sees a bug.
He measures it.
Its body is 1 inch.

Ann has a ruler. She measures her arm. It is 1 foot.

There are also centimeter rulers.

Kay
measures
a rose.
It is 15
centimeters
long.

Other TOOLS

You don't need a ruler to measure. Measure with your hand. Or use a crayon. What else can you use?

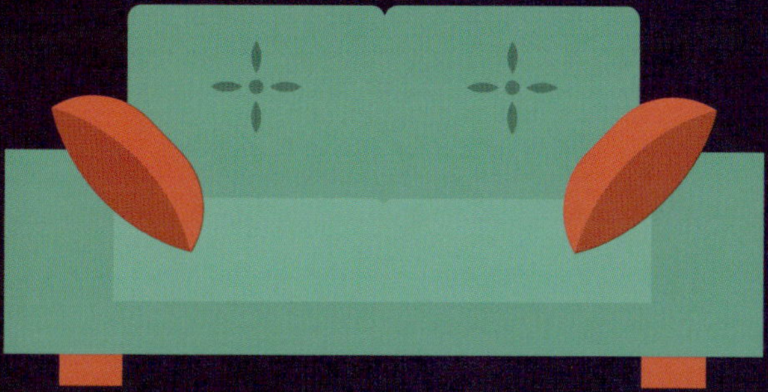

Try It!

Use this book to measure! How many books long is a table? A couch? A rug?

LONGER LENGTHS

A yard is 3 feet long.

A football field is 100 yards long.

Miles are very long.
There are 5,280 feet in a mile.

Some cities are miles apart.

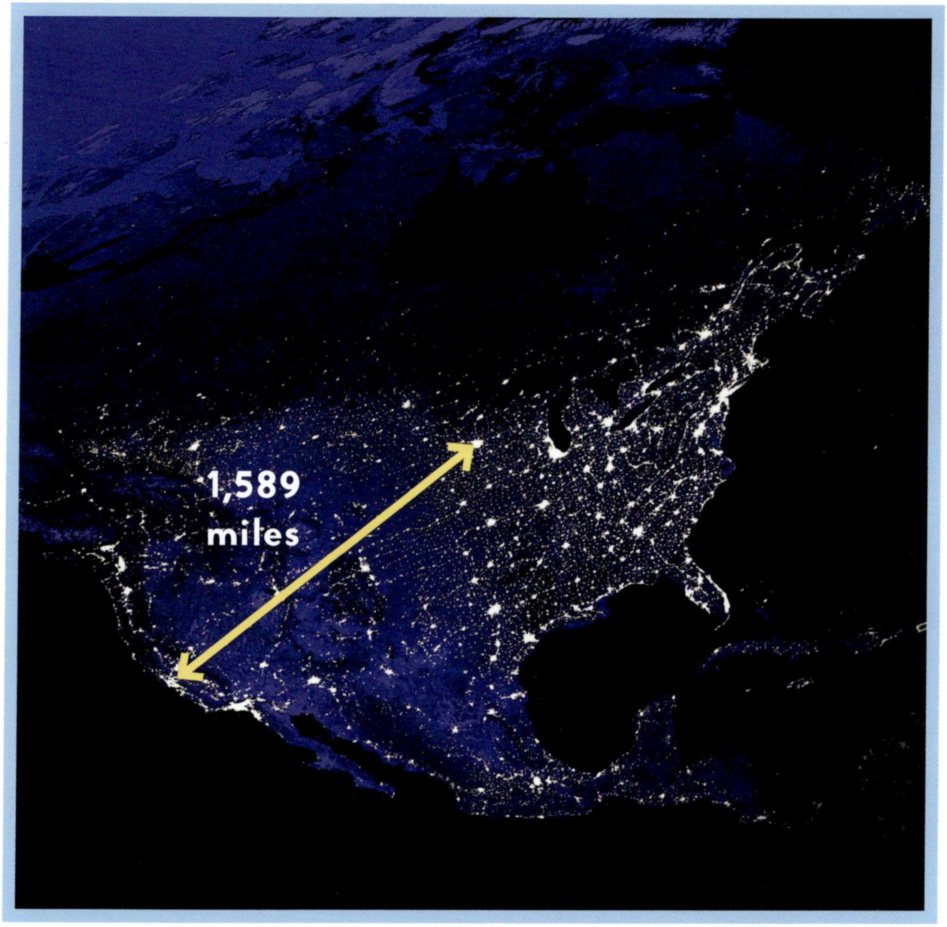

PRACTICE

Which picture shows measuring length? Which shows measuring distance?

1

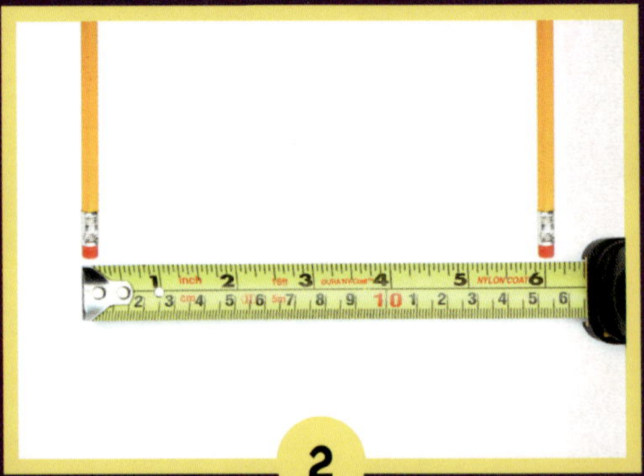

2

Answers: 1. length 2. distance

Look at the objects below. What is the best way to measure each one? Think about it. Then follow the lines. Were you right?

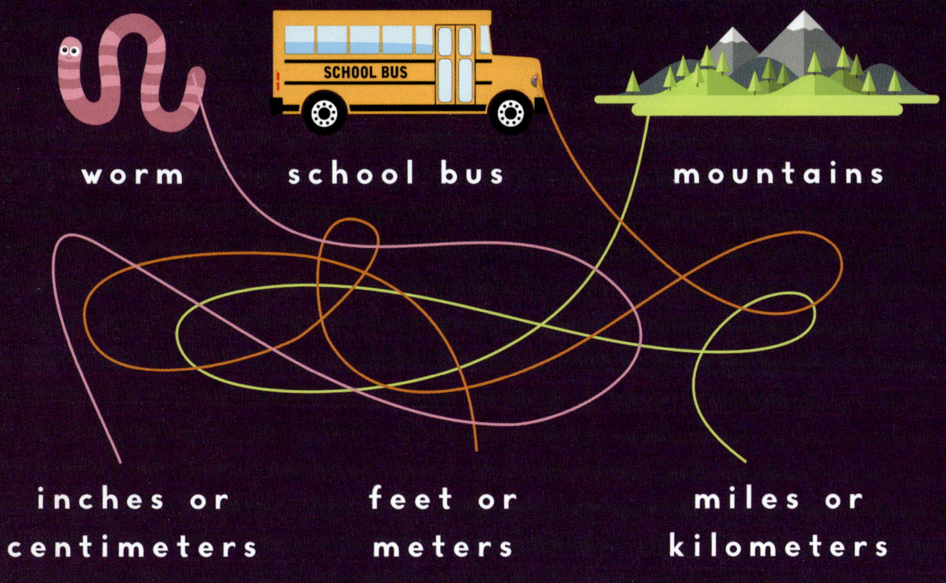

Glossary

APART – away from each other.

COUCH – a long piece of furniture that you can sit or lie on.

FIELD – an outdoor area where a sport is played.